Edward Forter Alexander

Railway Practice

Its principles and suggested reforms reviewed

Edward Porter Alexander

Railway Practice
Its principles and suggested reforms reviewed

ISBN/EAN: 9783337295684

Printed in Europe, USA, Canada, Australia, Japan

Cover: Foto ©berggeist007 / pixelio.de

More available books at **www.hansebooks.com**

RAILWAY PRACTICE

ITS PRINCIPLES AND SUGGESTED REFORMS
REVIEWED

BY

E. PORTER ALEXANDER

———

NEW YORK AND LONDON

G. P. PUTNAM'S SONS

The Knickerbocker Press

1887

RAILWAY PRACTICE.

The literature of that very considerable and important problem first called by Mr. Adams the Railway Problem, has recently been enlarged if not enriched by two proposed solutions. The one is offered by Mr. T. F. Hudson, in his volume called "The Railways and the Republic." The other by Prof. R. T. Ely, in *Harper's Magazine* for July, August, and September. A third proposed solution has long been before the public in the Reagan bill, which has twice passed the House but has failed in the Senate.

As the three solutions disagree radically in principle, and are at odds with the methods of reform which the railroad managers themselves are endeavoring to put into operation, it is clear that some confusion exists in the premises from which the different parties start. I propose to examine these premises briefly, and see if any such confusion cannot be removed.

Surely after fifty years of experiment, and a development covering all civilized countries, there must be to-day a few principles, settled by actual test, and put beyond question or dispute, making what we might call the present state of the science of railway management. If there are such, on which we can firmly ground ourselves, the discussion of the remedies for abuses becomes much more simple. I will take up singly the most important of the underlying questions which control the railway practice of to-day, and indicate and illustrate very briefly the conclusions upon them which have been reached, and recognized as final, by the universal practice of all countries in which railroads exist, whether operated by private parties or more or less under governmental control.

I cannot pretend to treat these questions exhaustively within my limits, but only to bring them clearly in the same field of view with the reforms proposed by Mr. Reagan, Mr. Hudson, and Mr. Ely. If my readers will accept what I shall thus lay down as "the state of the science," or its acknowledged principles, it will be easy to see how the new measures proposed square with them and with each other.

If my readers are not satisfied with my statement of these principles I can only assure them that the fault is due to partial presentation only, and commend any one desiring fuller information to the admirable work of Prof. A. T. Hadley, of Yale, on "Railroad Transportation," who has made a most exhaustive study of the railroad literature and practice of every nation, and presented it in admirable and condensed form.

COST OF SERVICE.

One of the principal points at issue between theoretical railway reformers and railway managers is, whether freight charges shall be based upon the *cost* of the service rendered, or upon its *value*.

Upon the answer to this question will depend the right or wrong of nearly all freight classification, and of most instances of charging less for a long haul than for a short. It is the universal custom among railroads the world over to base their charges upon the value of the service rendered, and not upon its cost—although the latter would seem to be the safer plan if they could only put it in force. It would seem as impossible for a railroad which could enforce this plan, ever to go into bankruptcy, as for the hotel in Arkansas, whose proprietor charged each guest the expenses of the house since the last one left, and collected with a shotgun. But railroads, in common with authors, doctors, inventors, laborers, lawyers, manufacturers, and most other people who have any thing to sell, base their prices upon the value of what they offer, rather than upon its cost. And indeed no other basis of price for railroad services is practicable, for the cost of rendering them is by no means the simple matter of calculation which it is often assumed to be. The cost of any particular act of transportation cannot even be averaged at, except under the most arbitrary assumptions. But few of any railroad's total expenses can be divided, and assigned to passenger, mail, express, and various kinds of freight, except by the vaguest guesswork. Results so arrived at would be as unreliable as the distance to the moon, estimated by measuring to the top of the highest mountain, and guessing at the rest.

The case of a railroad's estimating the cost of doing a particular piece of business is not unlike that of a lawyer estimating the cost of giving an opinion. He has fitted himself for that particular business, and, as it were, invested his life in the education and experience necessary to transact it. His time is good for nothing else, and, if he is not called upon for opinions, will be worthless to him. He can therefore render opinions up to a certain limit almost without cost, except for stationery. So a railroad is a large fixed investment capable of furnishing transportation and nothing else. Up to certain limits it can always take additional business without cost except for a very small amount of fuel. The money it receives for the new business above the small *additional cost*, is all clear profit. It adds that much to the ability of the road to serve other patrons at low rates.

It seems indeed to be unjust discrimination for a railroad to charge different rates for services that apparently cost it the same. It *is* discrimination, but when the *value* of the service is fairly considered, the injustice is but imaginary, and the results are beneficial even to those interests that seem to be discriminated against.

In another form this principle is sometimes expressed in the phrase that railway charges must be based upon " what the traffic will bear." This phrase has been as objectionable to some railway theorists as a red flag is supposed to be to a bull. Perhaps legitimately so, because the expression is vague and might be held to cover extortionate charges. As Prof. Hadley remarks, it has been interpreted to mean, " what the traffic will *not* bear." But in fact this expression is used, not to justify excessive charges, but rather to excuse the acceptance of rates which are lower than the average, and which thus appear to discriminate in favor of some particular interest.

Briefly, the principles upon which railway tariffs must be formed may be summarized as follows: They must not exceed a reasonable profit to the carrier as a maximum ; and may be reduced to what the traffic will bear as a minimum, if it does not involve the carrier in actual and permanent loss.

Between these limits they should be adjusted in proportion to value of service rendered.

It has been suggested that railroads should be forbidden by law to make rates so low as to yield no profit at all, or perhaps to involve a loss. Any legal restriction would seem to me very unwise, for two reasons. First, the point at which profit ends and loss begins cannot be definitely determined, because cost can never be accurately fixed, as explained before. Second, a traffic might yield a small direct loss, but indirectly a greater gain, by building up new interests, or sustaining old ones through periods of depression. Mines, mills, and manufactories often find it "profitable" to run temporarily at a loss, and railroads may do the same.

DISCRIMINATIONS.

When rates are based upon value of services rendered we necessarily have discriminations. In their infancy, railroads attempted to base their charges simply upon the ton-mile of service rendered.

Experience, however, and the demands and opportunities of rapidly expanding commerce and manufactures, not only taught but enforced the adoption of the value of service as the true basis of charges.

The same may be said, in fact, of the adoption of all the prominent features of modern railway practice. They have come about from necessity, not of choice, but actually against it. Like Topsy they are not creations but developments. Their adoption was under the constraint of laws which, whether understood or not, will always make themselves obeyed.

With the adoption of this basis for tariffs, it became possible for railroads to add enormously to the volume of their freights by transporting, in addition to ordinary merchandise, coal, stone, lumber, and many cheap and heavy articles which could not afford to pay average rates. Lower rates upon such articles were the earliest discriminations, but the results have justified the practice, and brought about in the end, (even for the high classes of merchandise which seem to be most discriminated against,) much lower rates than could ever have existed without the discrimination.

Such discriminations have resulted in freight classifications which grow more and more complex as traffic assumes larger proportions, and embraces more and more articles in competition with each other. As might be expected, classifications vary greatly in different countries, and often even in different sections of the same country, as they gradually force themselves upon railroad managers. Those in the United States are generally more complex than those in European countries.

It is doubtless in part due to this that our average

rates are lower than European rates, for in cost of materials and labor their roads can be operated more cheaply than ours, and they serve denser populations.

It is plainly a task of much delicacy and difficulty to adjust the comparative rates for the enormous variety of articles which must be transported. But the same commercial necessities which impose the task will guide to a gradual solution of it within reasonable limits.

The discriminations which have worked out such favorable results in freight traffic, produce equally favorable ones in passenger, as far as they can be applied. There are more difficulties in their application to passenger traffic in this country than there are in Europe. Consequently they have not been carried so far, and, as a result, our average passenger rates are higher than those of Europe.

The available methods of making such discriminations are by commutation tickets, and cars and trains of different classes. Their results have been most encouraging as far as tried, and there is still room for progress.

If it were only possible to apply these methods upon the street railroads in populous cities, it is evident that its results would be greatly beneficial, not only to the company but to the public.

If it were practicable to transport upon these cars, for one, two, or three cents, a large class of persons who cannot afford to pay five cents, the profits of the companies would soon be raised to a point which would enable them to reduce even their five-cent fares.

The public benefit of the lines, too, would be enor-

mously extended.   The prosperity which will attend
such an application of the principles of discrimination
will doubtless some day induce efforts to overcome the
difficulties in the way.

Closely akin to discrimination between articles is that of
discrimination between localities, or the giving of lower
rates to competitive points.   Discriminations in classification result from the fact that the same service has different values when rendered to different articles.   Discriminations between places result from the fact that the
same service has different values in different places.   This
difference is almost universally the result of natural
features or geographical locations.   The business of a railroad is simply the sale of *rapid* transportation.   Slow but
cheap transportation may be had between many localities by water, and still slower and more expensive transportation may be had by horses and wagons almost everywhere.   To those already enjoying water transportation,
transportation by rail offers comparatively few advantages,
and they can only afford to pay a small sum for it.   Nature has, so to speak, discriminated in their favor, and
given them what we may call natural transportation.   But
she has discriminated against inland places, and left them
dependent entirely upon artificial transportation, by horse
or man power, slow and expensive.   Hence the service
which the railway renders its inland customer is far more
indispensable and valuable to him than that rendered
those who enjoy natural transportation, and he can afford
to pay more for it.   In fact, the railroad is built, in general, only for the service of the inland party, and it sells,

as it were, its surplus power to the maritime party for any price it will bring. An example will illustrate: New York and San Francisco have always enjoyed water transportation, for freights from the one to the other, slow, but very cheap. No one would ever dream of building a railroad between those cities for the sake of the through business it could get in competition with the ocean. But between them lie wide stretches of lands against which nature has discriminated in the matter of transportation, while endowing them with great and varied wealth in agricultural, mineral, and other resources. Her discrimination was so heavy that only narrow margins of this vast territory could be utilized and developed with ordinary land transportation. To overcome this natural discrimination, railroads were built into the interior in every direction. With their gradual improvement in machinery and in methods of work, they have pushed into the remotest sections, and, making connections in each direction, at last we have through lines from New York to San Francisco.

As the railroads advanced, they enormously reduced the discriminations of nature throughout this inland territory. Thirty years ago it cost over a dollar a pound to carry from New York machinery and tools to work the mines of Utah, and the trip consumed the whole summer, during which the purchaser lost the use of his money. Now the trip requires but two weeks, or less, and the rate is about two cents. Comparing these rates, and considering the character of the present service as compared with the old, it is not an exaggeration to say that the railroads have removed about ninety-nine one-hundredths of the discrimination against Utah which nature ordained in surrounding her with deserts and mountains.

When the railroad connection is at last complete to San Francisco, the question arises at what rate will it take freight between San Francisco and New York? If the railroad manager could control the question, he would doubtless say that two cents being the rate from New York to Utah, about two thousand five hundred miles, at least two and a half cents should be charged to San Francisco—eight hundred miles farther. But San Francisco has always enjoyed water transportation, and can get her freight from New York by water at less than a half cent per pound. For increased dispatch, a shipper will pay the railroad perhaps a cent, or a cent and a quarter, on certain classes of freight. That price is then all that rapid transportation between New York and San Francisco is worth, and the railroad must sell at that price or not at all. If that price is more than the *additional outlay* involved in doing it, as against leaving it alone, it is profitable to the railroad, and the business is moreover advantageous to the whole inland community served by the railroad. For it adds to the number of men employed along the line, and contributes to the dividend and interest account of the railroad, and the more prosperous the road, the lower its local rates may be made.

Plainly the only limit at which the railroad should stop competing for through freight is at the *additional cash* outlay for doing it. That is the simple common-sense of the question. Railroads must and should compete at competing points down to a limit which is, not the average cost of doing all their business, but the extra cost of that special additional business.

We have illustrated by the case of New York and San Francisco, but the principle is the same in all other cases. and a fact not generally appreciated is, that it is the competition of the water routes that causes even the inland discriminations and competitions, so generally ascribed only to the folly and perversity of the railroads. A lowering of the rate on grain by lake and canal from Chicago to New York, for instance, will reduce the price of grain at Charleston and Savannah, which are supplied partly from New York by water and partly by direct rail lines from the west through Atlanta and Augusta. This direct rail line must then either lose all its Charleston business or accept lower rates upon it. This involves a disturbance of equilibrium between that city and interior markets which compete with it as a distributing centre. To protect their business the railroads are compelled to give competing rates to a greater or less extent to these interior distributing markets, and the disturbance rapidly spreads as far inland as the Ohio river. In fact, with our long railroad lines, the United States is no longer a continent, but a large island cut in two by the lakes and the Mississippi river. Our trunk lines have, practically, their termini upon the water at both ends, and most of their through business is done in competition with water routes.

A common fallacy in the discussion of this subject is the charge that the lower rates given to competitive points impose heavier burdens upon shippers at local and non-competitive points.

Should the rates, indeed, at competitive points be put below the " additional cost of new business " the railroad would lose money and be less able to afford low rates to its local points, but this state of affairs never prevails but in railroad wars, and is not to be considered as normal. Even in such wars I have never heard of a case where local rates were advanced to make good losses on the competitive business. It is often charged that this is done, but the charge is only on suspicion and without proof, and by theorists unfamiliar with railroad habits of thought and action. But, even should such injustice sometimes happen, the remedy is to be found in the checking of railroad wars, which, with their attendant evils, is to be considered later. Generally lower rates to competitive points not only benefit the local shipper in-directly, by adding to the prosperity of the railway which serves him, but really every reduction of rates to a com-petitive point is more a direct benefit to the surrounding local points, for which the competitive point serves as a distributing centre,.than it is to the latter itself.

For instance, farmers in the West receive for their grain a price which is practically the New York price less local freight from the farm to Chicago, and through freight from Chicago to New York. For their supplies from the East they pay New York prices plus through freight to Chicago and local to the farm. Now if the through rate from New York to Chicago is reduced, the farmer gets the full benefit equally with the Chicago shipper. The latter, in fact, is only the middleman whose function is the collection and handling in large quantities of the articles produced by and needed upon the farm in small quantities. He should simply pass the whole reduction in rate on to the farmer. If he does not it is not the fault of the railroad, except where rates arc allowed to fluctuate rapidly. Where rates remain steady the benefit of the low through rate, to the large distributive centres, is very certain to pass through the hands of the middleman, and reach the producer and consumer. Hence, as well as for many other reasons, the great desirability of steadiness and uniformity in rates.

SHORT VERSUS LONG HAUL.

There are very few, even among the purest theorists upon railway matters, who have failed to see the advantages accruing to all concerned from having railroads compete with water routes, and thus establish competitive points enjoying rates cheaper per mile than the locals. In other words, discrimination between places it is generally admitted must exist. But there is a limit set to such discrimination by many theoretical reformers which is very different from the limit I have above given. I have stated that, to secure new or competitive business, a railroad may, if necessary, reduce its rates down to the "additional cost of doing new business." But the limit of the theorists is the local rate charged to intermediate local points. If the through or competitive rate is reduced below that point, it becomes a case of the longer haul being done for the lower rate.

No railroad practice or abuse has ever brought out more violent and indignant protest than this. The boasted liberties of our country are pronounced a delusion and a sham while. such practices exist, and the railroad managers who make such rates are compared to the robber barons of the middle ages, or the Czar of Russia and oriental despots in general.

And it may be frankly admitted that to a superficial view the practice does seem to be at once arbitrary, unjust, and unnecessary.

But a consideration of the circumstances under which such discriminations arise will entirely change their aspect. They are not a wicked invention of railroad managers, devised to favor or to injure particular localities, but they are simply the inexorable results of geographical facts and the laws of trade, more powerfnl than railroad managers, czars, or despots.

I have shown above that water competition first gave birth to discrimination, giving lower rates to competitive points. *The competition which gives birth to such discriminations determines also their sizes, or the extent to which they must go.* What are the rates to intermediate points " has nothing to do with the case." If the competitive traffic cannot be obtained at, or will not bear, a rate as high as reasonable intermediate locals, it must be reduced below them. Actual figures will best illustrate, and it seems important to make this matter very simple, when it is so generally misunderstood that the Reagan bill gets an intelligent support largely because it forbids the practice under very heavy penalties. It seems to me impossible that any impartial mind can give it a careful study and fail to conclude that the apparent discrimination does not injure, but really benefits, the very places that seem to be discriminated against. A single case will serve as an illustration of all, for all result from similar causes.

I select, as an example, the rates upon a car load of sugar from New York to town and cities averaging about two hundred miles apart, along a line from New York to San Francisco via Chicago. See p. 16.

Table A gives such a list. Opposite each town is its distance from New York, and the rate per hundred pounds charged by the all-rail line upon shipments of sugar in car-load lots. As classifications are not uniform on all parts of the line, or between the rail line and water routes, it is necessary to select some single article for comparison. But for this purpose a single article is sufficient and sugar is selected as one largely transported. The rates given are those in force in July, 1886. Peace prevailed generally, and tariff rates were maintained everywhere but to the Pacific coast. Here, as is well known, for some months a severe war has been waged between the rail and water lines. The table shows the " tariff " rates to San Francisco which prevailed before the war broke out, when the rail lines pooled with the Pacific Mail and each other, and also the actual " war rates " offered at the time. I also give the Chicago rates by rail and lake, or one half water ; and by canal and lake, or all water. An important fact to note in this connection is that all localities west of Chicago share equally with that city in her low water rates, if they choose to use

TABLE A.

RATES ON SUGAR PER HUNDRED POUNDS IN CAR-LOAD LOTS.

From New York to	Dist. miles.	Rate, cents.	Graphic Comparison of Rates.
Harrisburg, Pa., rail	200	.15	
Altoona, " "	326	.15	
Pittsburg, " "	444	.15	
Bucyrus, Ohio, "	640	.21	
Hamlet, Ind., "	840	.25	
Chicago, Ill., "	942	.25	
" lake and rail		.20	
" lake and canal		.16	
Clinton, Iowa, rail	1080	.50	
Cedar Rapids, Ia. "	1162	.50	
Omaha, Neb., "	1432	.59	
North Platte, Neb. "	1722	1.19	
Cheyenne, Wyo., "	1950	1.54	
Green River, " "	2280	2.09	
Ogden, Utah, "	2466	2.14	
Elko, Nevada, "	2742	1.98	
Humboldt, " "	2926	1.98	
Sacramento, Cal., "	3210	1.25	
San Francisco, Cal.,			
Tariff, rail	3299	1.25	
War Rate, "	3299	.87	
Tariff by Isthmus		.60	
War Rate by Isth-			
mus		.35	
Tariff by Cape			
Horn		.40	
War Rate by Cape			
Horn		.27	

the water routes east of Chicago. The all-rail rate to Green River, for instance, $2.09 may be reduced 9 cents by having the freight shipped by canal and lake to Chicago for sixteen cents, instead of by rail for twenty-five cents.

This illustrates the statement heretofore made, that the low rates to distributing points do not end there, but are passed on to the local points beyond.

Looking now at the lines representing the rates, in the last column, the eye at once detects every case of charging less for the longer haul, by the line being shorter than some line above it, and the reason why the longer haul must be taken at lower rates is also plainly apparent.

Why are the rates for points west of Ogden less than the Ogden rates? Because if they were greater no freight would take the overland route for those points, but it would go around to San Francisco by the Isthmus, or Cape Horn, and come back eastward to its destination, by a short rail haul on the Central Pacific, for less money. Those points possess the natural advantage of being nearer the ocean than Ogden, and neither railroad official nor law of Congress can deprive them of it. And what must be the result if the Reagan bill is passed and the overland lines forbidden to take a less rate on the longer haul? They cannot advance the San Francisco rates, as they are controlled by the water routes; so they must either reduce the rates to all interior points or simply retire from the San Francisco business. Evidently they will take that horn of the dilemma which will least affect their *net* revenue.

Let us see which that will be. To reduce the offending rates between Omaha and Ogden to the San Francisco rates, would be to cut off the lines representing them by a vertical through the San Francisco rail rates as indicated. Or, in figures, when the railroad and the water lines are at peace, every rate above $1.25 would have to be reduced to that figure, and when they were in active competition, every interior rate would be reduced to 87 cents.

Now bear in mind that every cent of this reduction is a loss out of *net* revenue, for the expense of doing the business remains unchanged. Also, two other facts, not indicated by the table, but of great importance in determining the question. First, the amount of local business, upon which rates would have to be reduced, is probably three times as great as the through business in question ; Second, the net revenue on this through business, done at such low rates, is probably not a third of the net revenue which would be lost upon the local, pound for pound, by the reduction. Evidently the profits to the railroad would be as nine to one in favor of simply abandoning the entire through business, and confining itself to the local. It could not even force any compromise with the water routes, for the latter would appreciate that the railroads could not afford to enter into competition, but was bound hand and foot by the tying together of local and competitive rates.

Now let us see who would be benefited by this result. We will suppose that the law was passed at the request of Ogden, which felt itself outraged by having to pay

$2.14 for what San Francisco gets for $1.25 or $0.87. Ogden must still pay its $2.14, and the railroad which does its business will lose one ninth of its net revenue, and run one third less trains. Evidently Ogden is rather injured than benefited.

As to San Francisco, it must then get its freight entirely by Isthmus and sail. So that the only persons benefited are the carriers by water, between New York and San Francisco, who are relieved from rail competition, or have it handicapped with severe penalties. Should we join the ends of our lines representing rates in the table, we would have a line which we might call the curve of rates. It rises at it leaves navigable waters, and falls as it approaches them. There is an apparent exception in that there is no point higher than Chicago between that place and New York. There were such points in former tariffs, and during rate wars there will be again ; but the shortness of the distance, the proximity of lakes and rivers on each side of the line, the number of sharply competing railroads, the cheapness of coal and iron, and the dense population of the country have enabled the railroads to take them out when they are at peace and can maintain rates. But over long lines, across mountains and deserts, and in thinly settled countries the curve of rates will always rise over the interior. I think it safe to predict that, on the transcontinental lines, the curve will always retain its present general character, while transportation by land remains much more costly than by ocean.

Railroads may be forbidden by law, and under heavy

penalties, to work for such long-haul business, but the *rates* will be there all the same; and no one will be benefited by the railroad's exclusion from them but the Pacific Mail steamers and the clippers around the Horn.

Three mistaken ideas upon this subject are very prevalent, and have led to the popularity of "long- versus short-haul" legislation.

First.—That the railroads are losing money on the long hauls and making it up on the short; whereas, whatever is received for the long above "*additional cost*" is extra, and goes that far to help the road and all the country along it.   Second.—That only the city which receives the rate gets the benefit, when it is really but a benefit in trust on its way to local producers and consumers.   Third.—Every village believes that competitive rates would make a city of it, which question is one beyond discussion.

Finally, it may be remarked that the very phrase, "charging less for the long haul," in the essence of the matter, implies an error.   For the essence of the imagined injustice is in the idea that the long haul costs more than the short.   But, in the element of cost, six miles of ocean are scarcely equal to one mile of land, and, estimated by the cost, the distance from New York to San Francisco by water is scarcely as great as from New York to Omaha by rail.   As measured by cost of transportation by the shortest line, therefore, San Francisco is nearer New York than Ogden, and is entitled to the lower rate.

PERSONAL DISCRIMINATIONS.

I have shown above that discrimination between things, resulting in classification of freights, and discrimination between places, resulting in lower rates at competing points, even to the extent sometimes of charging less for the long haul, both result from the necessity that railroads are under, in common with private individuals who have services to sell, of basing their charges upon the value of their services.

I have shown also that the results of these discriminations are advantageous to the whole community, even to the localities which are apparently discriminated against ; but which would find natural or geographical discriminations pressing much more hardly upon them, should legislation interfere with the freedom of the railroads, the natural allies and friends of all interior towns, to compete for business and draw part of their revenues from distant territories and customers.

So far, then, as these discriminations are concerned, there is no railroad problem.

But there is another class of discriminations which have proven a problem indeed,—discrimination between individuals.

None of our railroads are able to command all the business their tracks will accommodate, and few of them find it easy to meet their fixed charges and dividends to stockholders. Consequently, the struggle for business is sharp wherever they come into competition, and the easiest way for any single road to get a great share of it is to pay the large shippers for it by giving them private rates or rebates. These rebates give the favored shippers undue advantages over their competitors in business, and are most cruel wrongs and injuries to the latter, and to the community as a whole. No language is too severe to apply to such discriminations, and none of the denunciations of Mr. Hudson or other writers on the subject are undeserved. The common law has always condemned them, and given damages to parties who could show themselves discriminated against.

But, unfortunately, there are peculiar difficulties in the way of preventing such abuses by legislation. The Reagan bill seems to be the best hope of those who have faith in legal measures, and certainly every one must approve its proposed allowance of exemplary damages, and its bringing legal relief more easily within reach of an injured party. But I do not believe that this or any other legal remedy can ever be effective. There are too many ways in which rebates may be hidden and covered, and the more severe the penalties the harder will be the proving of the crime. A mere verbal promise may be given and a private account kept, no money being paid until all danger of suspicion has passed, or until action would be barred by limitations. Or the railroad company can enter into other dealings, with a large shipper, to secure his business, buying goods from him at agreed prices, or lending him money without security. It is simply practically impossible to prevent a man's making a present to his neighbor if he chooses to do so, and it is equally practically impossible to abolish rebates by law. There is no objection to going as far as the law can go. There might even be added to the Reagan bill equal penalties for the receiver of a rebate as for the giver, but the only result would be that rebates would be more secret than before.

It must not be supposed, however, that railroad officials themselves are so enamored of this method of securing business that they do it from preference. It is an evil which arose very gradually, as long lines were formed and through business grew in magnitude, and it was only fully developed in the bitter wars which have both followed and fostered this growth. And, as there are compensations in all things, it may even have had its uses, heretofore, in teaching both railroads and the public how cheaply freights can be carried. But, however that may be, the railroad managers have not been behind the public in a most sincere desire to put an end to the system. A system of rebating can only be profitable to a railroad when its competitors are not resorting to it themselves. Then it becomes unprofitable to all alike, and it is not claiming any special virtue or honesty for them to assert the sincerity of the efforts they will make to break up the system. This leads us to the consideration of the subject of pools.

POOLS.

As the railroads had no legal restraint upon each other, and, from personal knowledge of the matter, appreciated that legal restraints could not reach the demands of the case, they have endeavored to find a remedy for the practice of rebating which would go to the very root of the evil.

The principle upon which they have based their remedy is that of *removing the temptation to give rebates.* It is certainly the most effective way of abolishing any crime, to remove the temptation to commit it.

Until ten or twelve years ago, agreements and promises between managers were relied upon entirely to secure uniformity and publicity of rates in competing business. About that time it became manifest that they could be relied upon no longer. It was not that any one could not confide in the personal good faith of any other one; but responsibility and power are necessarily greatly subdivided in the management of a railroad covering a thousand miles and fed by many connecting lines, and the manager must repose some confidence in subordinates who, as experts, have charge of various departments, and he must depend upon them for facts and information. Subordinates are always partisans, jealous of the rights of their particular roads, and suspicious of the fair play of others. Shippers, moreover, are too prone to believe that their best interests are promoted by having the railroads always at war with each other, and resort to all sorts of devices to destroy confidence between the railroads, and make each one believe that some other is giving rebates.

Where there are only two or three competitors, personal agreements to maintain rates may sometimes be sufficient, but where the number becomes great, confidence in the good faith of all is hard to be maintained.

When confidence is gone, the agreements are but ropes of sand, and it is as impossible to fix responsibility for this state of affairs upon any particular person, or railroad, as it would be to blame any particular member of Congress for its occasional waste of time upon trifling matters to the neglect of others of grave importance. It is simply the way that numbers affect joint operations. Methods that will suit small communities, or affairs upon a small scale, may be entirely unadapted to larger ones.

So the development of our railroad system, years ago, passed the point at which uniformity and stability of rates could be maintained by ordinary agreements, such as still prevail in most of the special industries of the country, and several years of bitter competition ensued which may well be characterized as wars, for the ruthlessness of their methods. It was during these wars that the Standard Oil Company acquired its first strength and prestige, and that the rebating system grew up, which was afterwards exposed by the Hepburn investigation. The Standard Oil Company for a long time received rebates from the railroads averaging over a half million per month, and in every trade and industry, which used the railroads largely, rates were honeycombed everywhere with rebates to those smart enough to get them.

The railroad situation then indeed became a problem, and to this problem there soon seemed to the railroad managers but one possible solution. And in all the writings and legislative debates which have gone on over it since, and among all the experiments which have been tried at home and abroad, nothing else has yet been suggested which has given any promise of success. Unre-

stricted competition, or war, among railroads, means dis-
tress and injury to the community. It means rebates to the
large shippers and a crushing out of the small. It means
fluctuating rates ; it means bankruptcy to railroads and
loss to investors, and finally financial crises and commercial panics, in which the poor and the weak suffer most.
And until some practical plan of restraining competition
within reasonable limits is devised, pooling must be re-
sorted to as the only refuge from immediate and greater
evils. It is not pretended that there may not be evils
and dangers inherent in the pooling system. But so far
as the present experience of the world teaches, it is
the only effective remedy, or palliation if one prefers,
for evils which cannot be at all times endured. But
it is not an easy remedy to apply. The difficulties of
establishing and carrying out a successful pooling
system are very great, and ten years of effort on the
part of the railroads have not succeeded in overcoming
them all.

One of the principal difficulties has been found in the
popular prejudice against them. The public has regarded
the pool as simply a device to advance rates to extortionate figures. A cry is raised against the " power " which
a pool is supposed to place in the hands of its managers.
Power is a very vague word until it is stated what work
the power is to accomplish. Applied to a railroad pool,
it can only mean a power to raise rates beyond reasonable figures, or to practise extortion.

Ten years ago theories upon this subject were in order,
for the experiment was untried. But now we have some

facts before which all theories must yield. We can now
at least judge of the *tendency* of the pool, and by comparing the present situation with that at its beginning,
can decide whether or not there is any palliation of
the personal discriminations from which deliverance
was sought, and whether or not there are any indica-
tions of the approaching extortion apprehended by the
public.

As to the decrease of personal discriminations, or the
prevalence of rebates, there are of course no statistics or
exact figures to be appealed to, but there are certain in-
dications from which the general situation may be in-
ferred. First, there is little recent public complaint of
this evil by shippers. Ten years ago the papers were
filled with it, and public meetings were frequently held
to denounce it. The indignation would not be less now
did it exist to an equal degree. The theorists who from
time to time write and declaim against the railroads,
draw all their examples from dates prior to the estab-
lishment of the pool, or from periods when its operations
were suspended. The oldest, and, in some respects, the
most successful pool in the United States, is that of the
Southern Railway and Steamship Association, which
covers most of the South Atlantic and Gulf States.
Before it was established the rates were honeycombed
with rebates at every competitive point in the territory.
To-day, from some personal knowledge of the situation
and from the best information I can obtain, I do not believe that there exists a single rebate or personal discrimination in all the pooled business in that territory.

As to the trunk-line pool in New York, it is believed by those most likely to know or to suspect such things, that none exist on the part of any of the trunk lines, though it is probable that some of the weaker connections still purchase a part of their business in some way.

If this is the case, it only illustrates the great difficulty in making the pooling system a success—the difficulty of disciplining those who do not adhere strictly to agreements.

Prof. Hadley writes on this subject as follows in his work, "Railroad Transportation," before referred to, page 249:

"The governments of Central Europe have given up trying to procure obedience to these principles by simple prohibitory laws, such as are occasionally proposed in Congress. They have a hundred times more police power than we have, but they do not undertake to do this. To secure obedience to this system, they must take away the temptation to violate it. This can only be done by a system of pooling contracts. These are accordingly legalized and enforced. They are carried on to an extent undreamed of in America. They have both traffic pools and money pools. There are pools between State roads and private roads, between railroads and water routes. It is regarded as a perfectly legal thing that one road should pay another a stated sum of money in consideration of the fact that the latter abstain from competing for the through traffic of the former."

It is certainly clear, then, that the pooling system at least greatly palliates the evils of discrimination and rebates. It only remains to see whether in avoiding these we are liable to have the public subjected to extortion. Here we may have exact figures to settle the question beyond all doubt, and I select a few examples to illustrate the rates before and after pooling.

From Chicago to New York fourth-class rates in January, 1876, were forty-five cents per one hundred pounds. In January, 1886, they were twenty-five cents.

But it is sometimes asserted that while the rates to competitive points, such as Chicago have declined, those to local points are maintained at exorbitant rates. I have therefore at random selected two rates, one partially a local rate, and the other entirely so, New York to Pittsburg, and Altoona to Pittsburg; and have made comparison with the rates of 1876. It is as follows:

Per 100 pounds.	Jan., 1876.	Jan., 1886.
New York to Pittsburg, 4th class	.30	.20
Altoona " " "	.28	.17

In the territory of the Southern Railway and Steamship Association a fair sample rate is that on compressed cotton from Atlanta to New York. At the formation of the pool in 1875 this rate was $1.10 per 100 pounds. It is now $0.75. A purely local rate from Loachapoka, Ala., to Montgomery, Ala., selected at random, shows cotton per 100 pounds, in 1875, $0.33; and in 1886, $0.24.

A graphic illustration of these rates makes a picture as suggestive as the patent medicine advertisements showing the patient before and after taking some great invigorator. Thus:

COMPARATIVE RATES.	BEFORE POOLING.	AFTER POOLING.
4th Class. Chicago to New York.	45	35
" " New York to Pittsburg.	50	20
" Pittsburg to Altoona.	35	17
Cotton. Atlanta to New York.	1.10	75
Leachapoka to Montgomery.	33	24

As to the effect of pooling upon rebates, as before explained, exact figures and data cannot be obtained, but a graphic illustration of it would be like the comparison of an enormous flock of crows with a few solitary blackbirds.

In the face of these facts, theorists like Mr. Hudson, who advances arguments based upon the report of the Hepburn Committee and other ancient history, are not entitled to attention. They have been left behind, and are threshing old straw. For the business portion of the community, those who use the railroads and pay them their freights, are rapidly absorbing the real facts of the situation.

Not only are rates lower, more uniform to all shippers than ever before, and less liable to fluctuations, but uniformity of classification is increasing, and harmonious arrangements between different lines are promoted, which greatly facilitate business.

An ideal system of transportation would be one in

which any shipper might sit quietly in his office, and contract to deliver freight at any town in the United States by referring to a printed tariff which would show rates as uniform to all as the rates of postage, and not exorbitant in amount. Let us see how near the existing state approaches the ideal, and whether its tendency is toward it or not.

The trunk-line railroads centering in New York now issue a little book of about forty pages, seven inches by three, which can be carried in the pocket, and gives the following information.

First, a classification of freight, embracing almost every thing known to commerce arranged in five classes.

It begins as follows:

PART OF TRUNK-LINE FREIGHT CLASSIFICATION.

Abbreviations used.—O. R.: Owner's risk. C. L.: Car Loads. L. C. L.: Less than Car Load. N. O. S.: Not otherwise specified. 2: : two times.

ARTICLES.	CLASS.
Acids, less than 50 carboys, O. R. breakage and leakage..2t	1
Acids, not less than 50 carboys, L. C. L., O. R.........	3
Acids, in barrels or iron drums, O. R...................	3
Acids, C. L., O. R.....................................	4
Acids, N. O. S., O. R..................................	1
Acid, Tartaric, in boxes or kegs.......................	2
Acid, Tartaric, in barrels or hogsheads................	3
Agate, Ware...	2
Agricultural Implements, N. O. S. (See Machinery.)....	
Alcohol. (Same as Liquids.)...........................	
Ale, packed in boxes..................................	1

Ale, packed in barrels or casks....................... 2
Ale, in wood.. 4
Alum, in boxes, kegs, or bags......................... 2
Alum, in barrels or casks............................. 4
Ammonia, dry, in boxes, kegs, or bags................. 2
Ammonia, dry, in barrels or casks..................... 4
Ammonia, liquid.   (Same as Acids.)..................
Anchors... 4
Animals. (See Live Stock)...........................
Antimony, metal, loose, in slabs or in boxes.......... 2
Antimony, metal, in barrels or casks.................. 3
Anvils.. 4
Apples, green, L. C. L., O. R......................... 1
Apples, green, C. L., O. R............................ 3
Apples, dried, in barrels or bags..................... 4
Apples, dried, in boxes............................... 2
Argols, in boxes, kegs, or bags....................... 2
Argols, in barrels or casks.......................... 4

Etc., Etc.

Next comes a list of all the important railroad stations
in the territory covered (which includes most of the
United States east of California and Nevada, except the
South Atlantic and Gulf States covered by the Southern
Railway and Steamship Association), and the rate on
each class of goods from New York to each point.

It begins as follows:

RATES OF FREIGHT.

FROM NEW YORK TO	PER 100 POUNDS.				
	1ST CLASS.	2D CLASS.	3D CLASS.	4TH CLASS.	SPECIAL.
Aberdeen, Miss.	$1 80	$1 54	$1 27	$1 02	$ 92
Ackley, Iowa	1 52	1 23	95	71	61
Ada, Ohio	64	51	39	30	22
Adams, Ind.	68	54	39	30	22
Adams Mills, Ohio	56	44	33	26	19
Addison, Ohio	61	49	38	29	21
Addison, Mich.	71	57	43	33	24
Akron, Ohio	53	43	32	25	18
Alamossa, Col.	4 90	3 99	3 22	2 53	2 42
Alanson, Mich.	1 05	85	65	50	40
Alba, Mich.	1 05	85	65	50	40
Albert Lea, Minn.	1 70	1 40	1 10	79	69
Albion, Pa.	53	43	32	25	18
Albuquerque, N. M.	4 52	3 89	3 20	2 64	2 53
Alexandria, Ohio	56	44	33	26	19
Alexandria, Ind.	69	55	41	32	23
Allegan, Mich.	75	60	45	35	25
Allegheny, Pa.	43	35	26	20	15
Alliance, Ohio	53	43	32	25	18
Altamont, Ills.	1 14	92	71	54	38
Alton, Ills.	87	70	52	41	29
Alton, Ohio	62	50	37	29	21
Almanda, Ohio	66	44	41	33	26

Lastly comes a list of about three hundred Eastern towns and manufacturing villages, from each of which the same rates apply as are given from New York.

It begins as follows:

OTHER POINTS TAKING NEW YORK RATES.

STATIONS. RAILROADS.

Abington, Conn...........................N. Y. & N. E.

Andover, Conn...........................N. Y. & N. E.

Ansonia, Conn...........................Naugatuck.

Anthony, R. I............................N. Y. & N. E.
Arnold's Mills, R. I......................N. Y. & N. E.
Ashuelot, N. H..........................Conn. River.
Auburn, Mass...........................N. Y. &. N. E.
Avon, Conn.............................N. H. & N.
Ayer Junction, Mass....................Wor. & Nash.
Baltic, Conn............................N. Y. & N. E.
Bartow, N. Y.........................N. Y., N. H., & H.
Bay Chester, N. Y....................N. Y., N. H., & H.
Bellingham, Mass......................N. Y. & N. E.

Etc., Etc.

Evidently the railroads appreciate what the country needs, and are striving to attain to it. It is of course possible that among a thousand kinds of freight from three hundred points of origin to twelve hundred points of destination, there may be many adjustments of rates and classification still to be made to secure perfect fairness, and such adjustments as are suggested by experience are constantly being made in conference between experts upon both sides.

With reference to one of them recently accomplished, the *Financial Chronicle* of August 14th said in its financial article :

"As illustrating the spirit of the times in the way of compromising difficulties and removing disagreements, we may cite the action of the trunk-line pool this week in yielding the demands of the dry-goods people for a lower classification for freight. Similar demands had been made before, when the pool was not so strong nor so firmly welded together, and when, therefore, the probability of granting the request seemed stronger, and yet the demand then was refused. Now, when the pool is on a very stable basis, and in position apparently to pursue an independent and arbitrary course, the efforts of the dry-goods people have met with a considerable amount of success. There is a lesson in this. It shows that the managers of the pool are neither obstinate nor unreasonable, and further, that they are not disposed to take undue advantage of the great powers possessed by them."

The same paper, in its issue of September 4th, has an editorial upon " The Effectiveness of Pools," from which I quote as follows :

" The old idea that a pool is a selfish, grasping monopoly, intent upon devouring every thing within its reach, desirous of stifling competition, and bent on levying exorbitant taxes upon the commerce and industries of the country, has given place to a much more rational and enlightened view. It is recognized now that it is a measure of self-protection, designed simply to avoid the evils of reckless competition.

"Enlightened self-interest has been the stimulating cause. There has been no desire to assume the aggressive as against other interests or other departments of industry, but rather an attempt to avoid self-destruction."

Finally I give, from Mr. Poor's excellent " Railroad Manual," the following figures showing actual results of the railroad operations of the United States for the year 1885 :

> 127,729 miles of railroad were represented by
> $3,817,700,000 stock,
> 3,765,700,000 bonds,
> 259,000,000 floating debt.

The gross earnings for the year were $772,569,000
Expenditures   "   "   "   "   503,075,000
Net earnings   "   "   "   "   269,494,000

The net earnings therefore averaged 3.5 per cent. upon the capital. Of the net earnings, $189,426,000 was paid as interest upon debt, averaging 4.77 per cent. upon the bonds, and $77,672,000 was paid as dividends, averaging 2.02 per cent. upon the stock.

Tonnage handled increased 10 per cent. over the previous year, but revenue received from it decreased 5.8 per cent.

The average freight rate per ton per mile was 1.057 cents for 1885, against 1.124 for 1884, and 1.24 for 1883, a decrease of about 20 per cent. within three years.

Surely there can be no charge of extortion made against a system showing such figures as the above. The car-mark of extortion is exorbitant profits to stockholders. When the stockholders receive such small returns, the rates as a whole cannot be exorbitant.

This record of our railroads is unexcelled, for cheap service and low returns upon capital employed, by that of any country in Europe, and is unequalled except, perhaps, by Belgium, though their dense populations, cheap labor and material, and low rates of interest give them many advantages in operating and maintaining their lines. In England the average returns on railroad capital are over 4 per cent. on both stock and bonds, which is not considered as low a rate there as it would be in the United States. In France the railroads pay about 4½ per cent. upon their bonds and 9 per cent. upon their stock.

Prussian (state) railroads pay about 5½ per cent.; Austrian, a little less. The Belgian system for many years paid 6 per cent., but has recently declined to 3, and is considered unprosperous.*

But it is asserted that much of the stock of our railroads is not legitimate, but is water. Such an argument may apply against any particular railroad that earns exorbitant dividends, but against the system as a whole it does not. For it would be easy to show that for every dollar of water in existing stocks, two dollars of the money of railroad investors has been lost like water spilt in the sand. Much of it was lost, doubtless, by bad judgment, but the fact remains that our existing system of railroads, as a whole, has cost fully as much as it is capitalized at. Scarcely one of them was originally built as it stands to-day. The earlier ones have been rebuilt and re-equipped three or four times, as experience pointed out necessary improvements. Many of them, too, were built before the business really demanded them, and the loss from this source has been enormous. Poor's "Manual" gives a list of railroads put in the hands of receivers during the year 1885, and it embraces 9,885 miles of railroad (nearly 8 per cent. of total mileage in the country), with $293,000,000 stock, $297,000,000 bonds, and $27,000,000 unsecured debt.

If the state would guarantee the interest upon money legitimately invested in railroad construction, investors would readily furnish all that might be desired, and rail-

* I am indebted to Prof. A. T. Hadley, of Yale College, for above facts concerning foreign roads.

roads could and would be built without watered stock.
But the state, very properly, refuses to assume any risk,
and leaves it to be borne entirely by the investor.   The
latter, then, having all the risk, naturally demands to have
also all the chances of profit if the road turns out a success.   He discounts the future and takes watered stock
to represent what he hopes will be his earnings.   That is
the only way that communities wanting railroads can
induce investors to supply the funds.   No other system
would suit American ideas and our form of government.
Some of the European governments have .adopted a
different course, and built and controlled in various ways
their own railroads.   But as a whole I think we need not
envy them their results.

But although our watered stocks are not at present the
basis of extortion in rates, nor very likely to become
so, there is another view of them in which it is not
so clear that public indifference to them should continue.   They are essentially like gambling inducements,
offered to investors to get their money into railroad construction.   It is as if the state allowed promoters of railroads to raise funds by means of lottery schemes, selling
as prizes the chances of growing rich upon future business represented by watered stocks.   So far it has been,
perhaps, a good thing financially for the state.   The
investors, as a whole, have had much of their money
swallowed up, and receive an average of only $3\frac{1}{2}$ per
cent. on the remainder, while the public has the railroads,
and very good ones too.

But I record my conviction that the practice of stock

watering should be prohibited, without much hope of ever seeing it done, and more on the ground that it is against public policy to make it easy for men to build railroads, or float any enterprises, with other people's money, than from the fear of railroads being enabled to practise extortion by the possession of watered stock. I think that at present the investor needs protection more than the shipper.

To return to the subject of pooling. I believe that the pool has come to stay. We have not yet learned how to best obviate all the troubles and difficulties attending it, and its operation is yet very far from perfect. But we are learning every day,—both the people and the railroads. With the fierce opposition of shippers everywhere at the commencement, it is a wonder that success has been so great as it has. But it has been fortunate that the experiment was made under a commissioner whose intelligence, experience, and integrity commanded the respect and confidence, even of those who differed with him most radically. At last he is winning for his work the recognition that it deserves—a work more important than any other which has been accomplished in the United States within twenty years. Mr. Fink is bringing about peace—peace in a war which unsettled all values, threatened the fortunes of the poor, and made all business extra hazardous.

But while the business men of the country are gradually learning to appreciate this, there is threatened still adverse legislation, designed to tear up and destroy all that has been accomplished in devising a successful pool, and to utterly prohibit the practice for the future.

Should it be enacted, there can be but one result. Railroads will not and can not forever fight. All wars must end in peace, and peace between rival railroads can only last when there is some community of interest. That may obtain in two ways—by a pool, or by consolidation under one ownership. Already great progress has been made under the second method, even while the pooling has been going on. The best that a pool can accomplish after all is but a partial community of interest, and where the rivalry is bitter consolidation is very apt to result.

Now, let pooling be forbidden or let it fail, and consolidation must be the inevitable result. It will become simply a commercial necessity, as resistless as the downward flow of the Mississippi to the sea. It may be temporarily checked in any manner that theorists think good, but it will have its way in the end.

PROPOSED PLANS OF REFORM.

I have endeavored above to give a clear idea of the underlying principles and the actual methods of the railway practice of to-day, and to show that they are essentially uniform in all countries, as preliminary to a brief discussion of the reforms suggested by Mr. J. F. Hudson and Prof. R. T. Ely.

Perhaps another preliminary should be some notice of the causes of their dissatisfaction with the present system.

Mr. Hudson's is very easy to understand. He is many years behind in the state of the science, and several in the history of the practice. He objects to any classification of freights ; he would limit competition with water lines; he would abolish all pools and prevent all consolidations. He believes that the state of affairs which existed ten years ago, as developed by the Hepburn Committee, before the formation of the trunk-line pool, still obtains, and his proposed reform is to return to the old theories which existed before there was any practice.

But Mr. Hudson is definite in his statements and easy to understand, and to answer if there be any reply; in which he is in strong contrast with Prof. Ely, in *Harper's Magazine*, before referred to.

The latter is an expert in political economy and scientific terminology, and his dissatisfaction, like one of Turner's paintings, while very evident in mass, is hard to lay hold on in detail.

A quotation will illustrate:

" Railways have perverted that normal and healthful dependence of man upon man which leads to the formation of the fraternal commonwealth,—a commonwealth of equal rights and privileges, such as our fathers aspired to found."

Now, I respectfully submit that to a practical man, face to face with the problem of trying to satisfy rival communities of merchants and shippers, employees asking higher wages, and creditors and investors seeking returns upon their capital, such criticism as the above is nonsense.

Does it mean that his railroad charges too much for carrying freight, and if so, what freight and where; or that he refuses to carry it at all ; or that he takes it and loses it on the way? If he carries it at reasonable rates, and delivers it all right at end of journey, what more is wanted?

What have our fathers to do with the case? But lest I
do Prof. Ely injustice, let me give a part of his more
formal arraignment and generous condemnation of every
thing about our railway system, and everybody con-
nected with it. He calls the latter " them" and all
other people " us," and declares that between " us " and
" them " there is no possible permanent relationship except that of " master and slave "—that " there is no middle
ground."

He formulates his charge as follows:

"I propose to show in these articles that our abominable
no-system of railways has brought the American people to a
condition of one-sided dependence upon corporations which
too often renders our nominal freedom illusory."

Again, I submit, the above means nothing. It is
like the poetry we read in our dreams, having rhythm
and harmony of sound, but very vague in its ideas. Who
did it? What did he do? Is it too late to stop him?
Will the tearing up of any old roads or the building of
any new tend to prevent our nominal freedom from
becoming illusory? If so, which roads?

From all the context I can get but one suggestion of
where the freedom of the people is restricted. and it lies
in the fact that shippers of freight are not free to name
the rates they will pay for the service, but that many
shippers have to ask rates from a single railroad.

The shippers being in numerical majority, he implies
that their wishes as to rates should prevail.

My limits will not permit more extended illustration of
what is nebulous in Mr. Ely's bill of complaint. I will only

say that when one who is before the public as an expert
and teacher in social and political economy, declares himself in favor of industrial and, if necessary, political revo-
lution, as Mr. Ely does, his language and charges should
not be vague and ambiguous. Other reformers are in the
field before him, advocating reforms and revolutions dif-
fering from his only in a degree and in the measures they
are prepared to use to carry them into effect. Indignant
declamation in sounding generalities furnishes aid and
comfort to the most visionary of socialists.

But the one matter of complaint upon which Professor
Ely is comprehensible, and is in accord with Mr. Hudson,
is the increasing size of railway corporations. Each re-
gards size as synonomous with power, and power they
assume to be only ability to do some great public evil,
the exact nature of which is not indicated.

It is true that our railroad corporations are growing
rapidly in size. So are many other corporations. It is
one of the developments of the age. With some this
tendency has caused a jealousy of all corporations. But
the corporation is the only means by which men of small
means can enter into large enterprises, and thus compete
with wealthy individuals. It is more than ever a commercial necessity.

And as large corporations can serve the public more
cheaply than small ones, they are learning to combine—
not only in the railroad field, but in many branches of
manufacture. Even in trade the large companies, with
the aid of cheap express and postal rates, are absorbing
the business which has supported heretofore small dealers
in many towns and villages.

But is the size of a railroad corporation synonomous with power to do public evil?

That is. the apprehension, but I think it is groundless. It is stated that with their wealth they are able to bribe legislators. Where legislators are corruptible it indeed scarcely requires great wealth to purchase them. But against the exercise of this power we are not without barriers. First is the great jealousy which exists toward railroads in the mind of the average voter. Second is the growing power of the press to concentrate a public opinion which even the most powerful corporation must feel. Third, our forefathers have set for us constitutional limitations which neither the bribed legislator nor the prejudiced voter can cross, which protect both railroads and people, each from the other, and under which there is no reason to doubt that both may continue to prosper as well in the future as in the past.

It is claimed that the legislature of Pennsylvania has belonged body and soul to the Pennsylvania Railroad; but if so, what have they done that is of public evil? So far as I can discover, their railroad legislation is confined principally to the mere granting of charters and powers to extend, consolidate, issue securities, and do such acts as railroads can only do by special authority. But there is nothing in the prosperity of the State, or the equal freedom and privileges of its citizens, to contrast unfavorably with any of the Western States, whose legislatures have been as notorious for hostility to railroad interests as that of Pennsylvania has been for subserviency.

If there is yet any single fact, any law passed by railroad influence in any state, threatening to liberty or oppressive of the people, I am unaware of it, and Mr. Ely and Mr. Hudson weaken their cases by not instancing it. Until such a basis of facts can be shown, I consider their predictions of the loss of our liberties by the power of railroads as groundless as they are vague.

I think the facts rather indicate that corrupt legislators are more apt to make a prey of rich corporations for purposes of blackmail than are such corporations likely to invade legislative halls with designs dangerous to liberty.

The recent case of the Broadway Surface Railroad Company will illustrate. For many years the great necessity for improved transit on this great thoroughfare in New York City remained unsupplied, until parties were found who would pay liberally for the charter.

Then only was the public want allowed to be supplied. The wrong was not in the thing done, but in the way of doing it. And the growing power of the press and of public opinion is exemplified in the punishment which has overtaken all concerned in it.

I do not believe, therefore, that the public is threatened with any such dangers as Mr. Hudson and Prof. Ely narrate. I think their fears are the result of entire misapprehension. But for all that let us examine their proposed solutions of the railway problem and see if they promise any thing better than we have at present. By way of criticism I need do little more than quote what each one says of the other's plan.

As among the learned men of Hindostan "who went to see the elephant though all of them were blind," those who respectively cling to the head and tail of the subject can refute each other's errors most effectually and without help from bystanders.

But first I should say that, were they practicable, neither of these plans would be at all objectionable to railroad stockholders and investors. My limits forbid my going into details, but either plan, if successfully carried out, would give to railroad securities something of the permanent and uniform character of government bonds. Rates of interest in many cases would be reduced, but the greater certainty of getting them would more than make up the loss. So it is only from the standpoint of the public that the proposed plans are to be criticised.

Prof. Ely's plan is simply that of some sort of governmental control. He does not commit himself to any detail, and he implies that a little reform of political morals and methods would be a preliminary necessity. He admits some difficulties and dangers, but finds ground for belief that all can be successfully overcome in two facts.

1. Our own government manages our postal affairs, and generally with honesty and efficiency.

2. The Prussian government controls the railway system of Prussia.

As to the first argument, I would say that there is such a radical difference between the relations to trade and commerce of the postal service and freight transportation, that no parallel can be drawn. Letters and papers have no commercial value. Postal rates are uniform for all distances. No commercial rivalries are affected by them, and little money is involved in the operations of the department but the salaries of the employes.

But in the transportation of freight, and the rates charged upon it, every branch of agriculture, manufactures, and commerce, with their infinitely diversified rivalries, is intensely, vitally interested. Rates cannot be uniform for all distances and nearly all articles, but there must be an infinite and frequently varying adjustment, or discrimination and adaptation of rates to circumstances.

An illustration will make this clearer.

With the invention and improvement of refrigerator cars a new business has recently sprung up—the shipping of dressed beef from the West. Eastern butchers who bring their cattle on the hoof and have heavy investments in stock-yards and abattoirs are threatened with destruction. Large sums of money and the livelihood of thousands of employees are involved in the comparative rates to be charged for dressed beef and live cattle.

With such questions the postal department has no concern whatever.

But suppose that a bureau of elected or selected officials had the fixing of those comparative rates. What a battle-ground would their offices be! What bribes would be offered, and when the terms of office expired what strenuous efforts by the disappointed party to change the officials!

This illustration is but one of thousands that might be given, and it will serve to illustrate another very important point to be borne in mind. It is that *there is no abstract principle of right or wrong, or honesty or dishonesty*, to which such questions can be referred.

My limits prevent my giving, by way of illustration, what can be said on each side of the dressed-beef question, but any intelligent person can doubtless imagine a good deal. In one respect the question is not unlike one which has recently vexed Congress—that between oleomargarine and butter. It is one between vested interests and new industries which encroach upon them. The bitterness of such contests and the extent to which partisans will go while engaged in them is well known. Usually the older industry will have the best of it if the matter can in any way be brought within range of governmental action.

Here the political strength of the older interest was able to drag it into Congress, which promptly strangled oleomargarine, the younger industry. Or a better metaphor would be to say Congress poisoned it. For under the constitution Congress has no power to strangle any infant industry, however repugnant its features may be to the older members of the family. But for purposes of revenue it has the right to prescribe a revenue tax upon any product. Although already embarrassed with surplus revenue, Congress deliberately prescribed an overdose of revenue tax for oleomargarine. No pretext was made that it was for any other purpose than to kill, and the strength of the dose was carefully estimated to produce that result. Now, however much we may all personally prefer butter, it is a very significant fact that the moneyed interest involved in this contest has led Congress to violate a plain provision of the Constitution. That has been done indirectly and by the perversion of powers given for other purposes, which it was openly admitted could not be done directly.

A later development of this matter is also interesting and instructive. From several quarters of the country come rumors that the oleomargarine interests have raised campaign funds and are in the field to secure the election of certain candidates for Congress or the defeat of certain other candidates.

Will not some modern George Stephenson arise and say, "the state must control oleomargarine or oleomargarine will control the state!" Or should the aphorism relate rather to butter? But the moral of all seems to me very plain. It is no small task to keep our political administration pure even when its activities and duties are limited to the most essential functions, such as preserving the peace, transporting the mails, providing for national defence, and such matters of vital necessity for the body politic. But that task will become utterly hopeless whenever Congress, a bureau, or any governmental agency is charged with the decision of the questions involved in commercial rivalries, such as that between dressed beef and live cattle. Many of these questions too, it must be noted, tend to bring about sectional issues, such as those involving the differential rates to be charged to and from Boston, New York, Philadelphia, Baltimore, Norfolk, Charleston, etc., and the competing points in the interior.

But my limits will not allow further remark upon this line, as I wish to notice very briefly Prof. Ely's reference to the Prussian governmental railroad management.

I do not profess to know a great deal about it further than that the geographical, political, and commercial situation is very different from our own. Prussia is small and finished. We are very large and growing. There it is a recognized governmental duty to look after and take care of the people. Here the people take care of themselves, and it is our theory that free competition will work out the best results possible to human nature.

I recently heard, incidentally, a single illustration of the workings of the Prussian system, which seems to me to indicate that it would be ill adapted to our wants.

About 1883, the wire manufacturers of Westphalia applied for a reduction of certain rates which placed them at a disadvantage in competition with manufacturers elsewhere. This request went up through regular channels, and commissions were appointed, and reports made and submitted to higher authorities, and counter arguments were made by rival manufacturers, and testimony was taken, and every thing was printed in a large volume. After two years consumed by all this the request for lower rates was refused.

There is however one part of the subject which Mr. Ely has apprehended very clearly, and which I refer to as bearing upon what I have said upon the subject of consolidations and poolings. He recognizes that only by means of combination and concentration can the transportation business of the country be conducted effectively and cheaply, and, instancing the purchase of the West Shore Railway by the New York Central, he states that the indignation of the press and public over this consolidation is more foolish than that of the laboring men who resist the introduction of improved machinery. He says:

" The impulse to such great economies as can be secured by
combination is irresistible.   It is one of those forces which
overwhelm the man who puts himself against them."

But upon Mr. Ely's general proposition of governmental control I can submit no better comments than
those of Mr. Hudson :

"A conclusive argument against the operation of railways by
the state is that it would introduce into our politics a vast
amount of patronage, which must largely become the spoil of
professional politicians.

"If the present business of the government, involving the
collection and distribution of about $300,000,000, cannot be
administered with a view solely to efficiency and economy, but
is used as the reward of skill in swaying the popular vote, any
reformation of our politics would be made utterly helpless
if that patronage were increased by an interest involving
$900,000,000 additional gross revenue, controlling nearly
$4,000,000,000 of property, and exercising a power over the
business interests of the country, beside which that of political parties is now trifling.   When our politics are purified
so as to exclude from them sefish ends and improper means,
it may be possible to bring the railways under political control
without making them a source of general corruption.   But
when such a millennial stage in human progress is reached
there will be no need of railway or other reform.

"At present, whenever railway management is closely con-
nected with politics, it leads to bribery, manipulation, and be-
trayal of public trust.

"All means of corruption would be multiplied by making
that connection universal and permanent, and converting the

control of the railways into the prize and sustenance of politicians and wire-workers.

"Apart from the danger to public morals of a state railway system, what guaranty is there that it would act with more impartiality and justice than private corporations ? Why should political control be more unselfish and equitable in managing transportation than private control ?"

Mr. Hudson's theory of reform is exactly the opposite of Mr. Ely's. He proposes to take away from the railroads their engines and cars, and leave them simply the right to maintain their tracks, and charge toll to a multitude of small private carriers who would each own his own rolling stock. Governmental commissions would regulate the tolls each road should be allowed to charge, and Mr. Hudson makes rough calculations what such figures might be. They vary from 0.189 cent per ton-mile upon the Pennsylvania, and 0.431 upon the Erie, to 2.42 upon the Denver and Rio Grande. It seems incredible that, having gone this far in his calculation, he did not perceive that with such a difference of rate between the Erie and the Pennsylvania, the through business would all immediately desert the Erie and go to the latter road.

This increased traffic would enable the Pennsylvania to make a further reduction of tolls, and the decrease on the Erie would force it to charge still higher rates in order to maintain its track in order even for local business. But it is unnecessary to follow out the obvious complications which would ensue from any attempt to base a transportation system for the country upon a system of tolls for the use of highway.

Professor Ely comments upon Mr. Hudson's plan as follows :

" The second plan is that of Mr. Hudson—the separation of railway ownership from the business of transportation.

" The idea is that the railway should become a public high-way, and should be compelled to allow all persons to run freight and passenger cars with locomotives over it, provided only that these be placed under proper supervision to prevent accident, and that a reasonable toll be paid. This is the theory that obtained universally in the first days of the rail-way, and it is doubtless this theory that led to the decision in Holland to intrust to the state the construction of the railway, and to allow private parties to take charge of the transportation of passengers and goods. Perhaps the strongest plea in favor of this theory is to be found in Mr. Hudson's book, but no in-stance can be adduced of any practical success in the applica-tion of this proposed method, and it is difficult to see any thing further in this restatement of the arguments for a separation between the ownership of the highway and the business of a common carrier than the lengths to which an able man may be driven who once determines to adhere at all costs to the doctrines of universal competition.

*　　*　　*　　*　　*　　*　　*

" Why is there not now competition in the express business ? We observe a monopoly, express companies having divided territory, although this does not at all seem to be due to the character of our laws. Competition has been attempted be-tween Baltimore and New York, but the Adams Express Company has crushed it. An old-established company, with wide ramifications and large capital, will even do business for nothing, between two main points, for a time, to ruin an obnoxious rival, and will maintain its life from other revenues, and look to a free field in the future for profits. If the separation suggested was effected, what guaranty have we that similar phenomena in the transportation of freight would not manifest themselves? . Again, there is great economy and convenience in the conduct of the transportation of freight and passengers by those operating on a vast scale, whether they own the tracks or not, and this gives to that industry its inherent and irresistible impulse toward monopoly, and, as already remarked, we desire these advantages. It is not clear that the technical difficulties of railway management do not interpose other and insuperable obstacles to the plan proposed by Mr. Hudson."

It seems unnecessary to say more upon this head except as to the technical difficulties referred to in Mr. Ely's last sentence quoted above. One is of such magnitude that it is surprising that Mr. Hudson overlooked it. It might be possible, indeed, that the railroad company might not only provide the track, but maintain a schedule for trains, and a train-dispatcher's office, which would prevent collisions and interference of the trains of different carriers while on their journeys. It could scarcely be *as* safe as the present system, but we may, for sake of the argument, imagine difficulties of that kind overcome. But it must ever remain impossible for many rival carriers to *occupy the same depots and make up their trains with shifting engines in the same yards.* Each carrier must have his own yards and terminal facilities. But available space for such facilities in our large cities can only be had at enormous expense. A few large transportation companies would speedily be formed, who would monopolize the entire transportation business of the country. Small carriers could not possibly gain or maintain a footing against them. And the large companies would speedily unite, and pool or divide territory. If Mr. Hudson had set out to devise a plan by which the transportation interests of the United States could be most rapidly consolidated into the most complete and irresponsible monopoly possible, he could not have suggested any thing half so certain and speedy of operation as what he has suggested to bring about the very opposite result.

CONCLUSION.

I have spoken so far of the railway problem as involving only the principles upon which tariffs are formed and competition between water lines and rival railroads is conducted.

But there is a second and a very different problem, which is confused with this in the minds of the public and of many writers.

This second problem is purely a social one. I do not propose to discuss it, but only to set it clearly by itself apart from the railway problem, with which it should have no connection.

It is the question, what is to be done, if any thing, about the accumulation of very large individual fortunes. It is commonly assumed that these fortunes are drawn principally from the community of producers and consumers by transportation charges, and accumulated by the lucky owners of railroad stock. In reality there are perhaps fewer fortunes so accumulated than are gained on the average in trade or manufactures. But the great majority of the phenomenal fortunes of the day are the result of what may be called lucky gambling. I do not mean to imply any necessary moral obliquity by the term gambling, for it would be a difficult task to draw the line between what is called speculating and is considered respectable, and what is disreputable as gambling. But speculating implies a slower mode of procedure than prevails nowadays with our exchanges, our network of wires, our tickers, and our general business rush. Man is a gambling animal by nature, and modern methods have enormously developed both its facilities and temptations, and have opened large fields in which gambling is not held to be disreptuable. Under such stimulus is it wonderful that its growth has been phenomenal ? Wall Street is its headquarters, and millions upon millions of dollars are accumulated there to meet the wants of the players. It has become the centre of the financial affairs of the country, as it is always ready to employ any idle capital. Railroad stocks are its favorite cards to bet upon, for their valuables are liable to constant fluctuations, on account of weather, crops, new combinations, wars, strikes, deaths, and legislation. They can also easily be affected by personal manipulations. Other things are dealt with as well, such as cotton, grain, petroleum, metals, etc., but the more active cards and the more rapid fluctuations are the most popular.

In such tremendous gambling large gains necessarily come to the lucky by the simple theory of chances. And when once a man is lucky enough to secure a considerable sum he is not so easily wiped out by temporary misfortune as the new-comer, and with larger stakes he can play safer games. And so the fortunes started by luck afterward grow by the inherent and attractive power of money. But the money which composes them is *the money won from the unlucky, and not the money, or in very small part that, earned by the railroads for transportation.* More and more every year are men, from all parts of the country, taking their surplus earnings in trade, in manufactures, in farming, and in all their multifarious pursuits, and bringing them into Wall Street to bet upon railroad cards. There is but one consolation about it : no man is obliged to contribute to these fortunes who does not choose to risk it. However alluring it may be, the man who resists its temptations can always keep his own money.

Our forefathers who framed the Constitution evidently thought the accumulation of very great individual fortunes an evil to be guarded against, and they took action to prevent it by forbidding the entailing of estates beyond the second generation. In those days it was imagined that even in two generations no excessive fortune could be accumulated. But our modern methods have made a very different world from what then existed.

Fortunes are now made in a few years which could scarcely have been accumulated in generations a century ago. Even the Rothschilds are now said to be nearly equalled by men who, thirty years ago, were poor, but who have been lucky in enormous speculations.

Money makes money, and money in great masses has its attractive power increased.

The aspect of phenomenal fortunes, therefore, is a social problem of some importance. Their manner of growth and their manner of use are to be observed, and what restrictions, if any, should be placed upon their accumulation are to be considered.

But I only desire here to draw the line sharply between these questions and those involved in the making of tariffs and conducting competition. The latter, indeed, are hardly any longer to be called questions.

Wherever in the world railroad transportation is conducted upon a large scale, commercial necessities have enforced the adoption of the principles I have endeavored to make clear, and their adoption has everywhere been marked by better, cheaper, and more uniform service. These universal features of railway practice may in fact be called its "adaptations to environment." They are as sure to assert themselves as the laws under which water crystallizes in fixed shapes, wherever and however it may be frozen. They are essential features of any transportation service adequate to the present needs of commerce. I will sum them up briefly :

Railway tariffs must be based upon value of service rendered, and limited by a reasonable profit upon cost of service and investment employed. In other words, they may be said to be bounded above by the line of reasonable profits, and below by "what the traffic will bear."

This lower limit, in adjusting itself to innumerable conditions produces discrimination between things and dis-

crimination between places, even to the extent of sometimes giving the lowest rate to the longest haul.

Legislation against such discrimination is legislation
against the competition which produces it—usually the
competition of rail lines with water lines.

Discrimination between individuals similarly situated
is unjustifiable and harmful.

It is so easily concealed that direct legislation against
it is not effective.

It can only be abolished by removing the temptation to
commit it. This can only be done by division of territory, by pooling, or by consolidation.

Division of territory is scarcely practicable in the United
States.

The commerce of the country could not exist under a
multitude of small carriers. Among such, instability and
uncertainty of rates and the existence of rebates would
be perpetual. A merchant desiring to ship goods would
have to go shopping for his rates as ladies do for their
bonnets. The rates would vary in different shops, and
from day to day in the same shop. Stability, publicity,
and uniformity of rates can only exist where there is a
community of interest between the carriers. This community can only exist under a pool or under consolidation.

Pooling has had a measure of success sufficient to war-
rant some confidence and further trial. The apprehension
that either pooling or consolidation results in extortionate
rates is conclusively contradicted by results.

Consolidations promote economy and efficiency. Their
political power is limited by jealousy of the press and

popular prejudices.　At best, its field of action is confined to procuring charters, powers to issue securities, and such privileges incidental to its business as can only be acquired by legislation.　It is more apt to be preyed upon than to prey.

And, finally, there are constitutional limitations upon the power both of railroad wealth and of popular prejudices, upon which public confidence may repose calmly. Hedged in and compelled by these barriers, the people and the railroads are daily learning the lesson that their interests are identical, and the prosperity of the one is inseparably bound up with that of the other.　The relation between them of "master and slave," advocated by Prof. Ely, would be one of universal disaster, and is impossible of permanent existence.　Prophets of such evil should be ranked with those who predict earthquakes and tempests from the planetary aspects.　Our plan of government is no failure, our fathers built not badly or weakly, and under the free operation of commercial laws we are to-day leading the world in cheap and efficient transportation as we are in the general prosperity inseparable from it.　There are, of course, inequalities in human lots and hardships inseparable from that great struggle for existence through which it is ordained that all development of better results shall come.　But we can hopefully await such development.

ECONOMIC AND POLITICAL SCIENCE.
RECENT IMPORTANT PUBLICATIONS.

Practical Economics. A collection of Essays respecting certain of the Economic Experiences of the United States. By DAVID A. WELLS, 8vo, cloth $1 50

CHIEF CONTENTS.—A Modern Financial Utopia—The True Story of the Leaden Images—The Taxation of Distilled Spirits—Recent Phases of the Tariff Question—Tariff Revision—The Pauper-Labor Argument—The Silver Question—Measures of Value—The Production and Distribution of Wealth.

"In my clear opinion, it is the most comprehensive, conclusive, and powerful statement of the truth respecting freedom of exchange, as to theory and as to practice, that exists in any language or literature."—MANTON MARBLE.

The Industrial Situation and the Question of Wages. A Study in Social Physiology. By J. SCHOENHOF, author of "Destructive Influence of the Tariff," etc. Questions of the Day Series, No. XXX. 8vo, cloth $1 00

The Destructive Influence of the Tariff upon Manufacture and Commerce, and the Facts and Figures Relating Thereto. By J. SCHOENHOF. Questions of the Day Series, No. IX. 8vo, cloth, 75 cents; paper 40

"An able presentation of the subject by a practical man, which should have a wide circulation."

Protection to Young Industries, as Applied in the United States. A Study in Economic History. By F. W. TAUSSIG, Ph.D. Questions of the Day Series, No. XI. 8vo, cloth 75

"There is a great deal of curious and important information condensed into this little book."—*Saturday Review*, London.

"His argument is entirely of an historical nature, and he presents many valuable facts and figures."—*Evening Wisconsin.*

The History of the Present Tariff, 1860-1883. By F. W. TAUSSIG, Ph.D. Questions of the Day Series, No. XIX. 8vo, cloth . 75

"Tracts like this will be read by many who would not open a bulky volume of the same title, and they will find that what they regarded as the most confused and perplexing of subjects, is not only comprehensible but also interesting."—*The Nation.*

The Distribution of Products; or, The Mechanism and the Metaphysics of Exchange. Three Essays. What Makes the Rate of Wages? What is a Bank? The Railway, the Farmer, and the Public. By EDWARD ATKINSON. Second edition, revised and enlarged, with new statistical material. 8vo, cloth $1 50

"It would be difficult to mention another book that gives so effective a presentation of the present conditions and methods of industry and of the marvels that have been wrought in the arts of production and transportation during the past fifty years."—*Advertiser*, Boston.

The History of the Surplus Revenue of 1837. Being an account of its origin, its distribution among the States, and the uses to which it was applied. By EDWARD G. BOURNE, B.A. Questions of the Day Series, No. XXIV. 8vo, cloth $1 25

" We especially commend this monograph to the consideration of those who have been captivated by Mr. George's schemes, as showing what usually occurs when governments undertake to expend large sums for the benefit of the public."— *The Nation.*

Postulates of English Political Economy. By the late WALTER BAGE-HOT, with an introduction by Prof ALFRED MARSHALL, of Cambridge, England. Cloth $1 00

" It will be especially valuable to those who have already some elementary knowledge of political economy ; for it makes more clear to them the true meaning of the doctrine laid down broadly in the elementary books, and will suggest the qualifications which must be attached to those doctrines."—Prof. F. W. TAUSSIG, Harvard University.

Railroad Transportation ; Its History and Its Laws. By ARTHUR T. HADLEY, Commissioner of Labor Statistics of the State of Connecticut, and Instructor of Political Science in Yale College. 8vo, cloth,

$1 50

" This book deals with those questions of railroad history and management which have become matter of public concern. It aims to do two things —first, to present clearly the more important facts of American railroad business, and explain the principles involved ; second, to compare railroad legislation of different countries and the results achieved."—*Author's Preface.*

The American Caucus System ; Its Origin, Purpose, and Utility. By GEORGE W. LAWTON. Questions of the Day Series, No. XXV. 8vo, cloth $1 00

" The author of this learned and interesting treatise never loses sight of his prime aim, which is to convince the reader that he may mend, but cannot suppress the caucus."— *The American,* Baltimore.

The Origin of Republican Form of Government in the United States. By OSCAR S. STRAUS. 12mo, cloth extra $1 00

" The undertaking was one well worth the labor involved, and its accomplishment is such as the historian may be congratulated upon."—*Transcript,* Boston.

The Science of Business. A study of the Principles controlling the laws of exchange. By RODERICK H. SMITH, 8vo, with charts, $1 25

" Eminently practical . . . Of interest both to students of economics and to business men."—*Am. Journ. Education.*

G. P. PUTNAM'S SONS, Publishers,

NEW YORK:
27 and 29 West 23d St.

LONDON:
27 King William St. Strand

www.ingramcontent.com/pod-product-compliance
Lightning Source LLC
Chambersburg PA
CBHW021524090426
42739CB00007B/770